Do It Yourself
HOMESCHOOL
JOURNALS

Copyright Information

Contact Us:

The Thinking Tree LLC

617 N. Swope St. Greenfield, IN 46140. United States

317.622.8852 PHONE (Dial +1 outside of the USA) 267.712.7889 FAX

www.DyslexiaGames.com

jbrown@DyslexiaGames.com